Dr. Birdley Teaches Science!
Classifying Cells

Featuring the Comic Strip

Middle and High School

Innovative Resources for the Science Classroom

Written and Illustrated by Nevin Katz

Incentive Publications, Inc.
Nashville, Tennessee

About the Author

Nevin Katz is a teacher and curriculum developer who lives in Amherst, Massachusetts with his wife Melissa and son Jeremy.

Nevin majored in Biology at Swarthmore College and went on to earn his Master's in Education at the Harvard Graduate School of Education. He began developing curriculum as a student teacher in Roxbury, Massachusetts.

"Mr. Katz" has been teaching science over 6 years, in grades 6 through 11. He currently teaches Biology, Environmental Science, and Physical Science at Ludlow High School in Ludlow, Massachusetts.

Nevin's journey with Dr. Birdley and the cast began in the summer of 2002, when he started authoring the cartoon and using it in his science classes. From there, he developed the cartoon strip, characters, and curriculum materials. After designing and implementing the materials, he decided to develop them further and organize them into a series of books.

Cover by Geoffrey Brittingham
Edited by Jill Norris
Copyedited by K. Noel Freitas and Scott Norris

ISBN 978-0-86530-539-7

Copyright ©2007, 2010 by Incentive Publications, Inc., Nashville, TN. All Rights Reserved. The *Dr. Birdley* comic strip and all characters depicted in the comic strips, Copyright ©2007 by Nevin Katz. All rights reserved. The Dr. Birdley logo, Dr. Birdley™, Clarissa Birdley™, Jaykes™, Dean Owelle™, Professor Brockley™, Gina Sparrow™, and all prominent characters featured in this publication are trademarks of Nevin Katz.

No part of this publication may be reproduced, stored in a retrieval system, or transmitted in any form or by any means (electronic, mechanical, photocopying, or otherwise) without written permission from Incentive Publications, with the exception below:

Pages labeled with the statement **©Incentive Publications, Inc., Nashville, TN** are intended for reproduction within the purchaser's classes. Permission is hereby granted to the purchaser of one copy of **CLASSIFYING CELLS** to reproduce these pages in sufficient quantities for meeting the purchaser's classroom needs only. Please include the copyright information at the bottom of each page on all copies.

2   3   4   5   6   7   8   9   10   10

Printed by Sheridan Books, Inc., Chelsea, Michigan • August 2010
www.incentivepublications.com

TABLE OF CONTENTS

Objectives & Frameworks .... 2 and 3

Overview of Source Cartoons 4

Teacher's Guide 5

Unit 1: Body Cells 13

Unit 2: From a Cell's Point of View 25

Unit 3: Cells and the Kingdoms 33

Unit 4: Two Types of Cells 45

Unit 5 and 6: Origins and Temperature Tolerance 55

Unit 7: Cell Scientists 67

Answer Key 88

Educational Objectives

Central Goal:
- To discuss major types of cells
- Discuss how they were discovered
- Introduce scientists in cell biology throughout history.

Chapter	Primary Objective(s)	Standards
1. Body Cells	1. Show that cells are alive using several body cells as examples. 2. Introduce the appearances and functions of different body cells.	3
2. From a Cell's Point of View	3. Distinguish between multicellular and unicellular organisms.	3, 4, 5, 8
3. Cells and the Kingdoms	4. Identify the major kingdoms of life. 5. Classify multicellular and unicellular organisms by kingdom.	1
4. Two Types of Cells	6. Discuss the major characteristics of prokaryotic and eukaryotic cells.	1
5. Origins	7. Illustrate the origin of prokaryotes and eukaryotes. 8. Compare the advantages of the two cell types.	8, 2
6. Temperature Tolerance	9. Introduce an advantage that prokaryotes have over eukaryotes.	8
7. Cell Scientists	10. Introduce the contributions of five scientists who contributed to the discovery of cells and the formation of the cell theory.	6, 7

FIND OUT WHICH CHAPTERS MATCH UP WITH YOUR LESSON GOALS!

Relevant Frameworks

This page outlines the relevant national frameworks that the chapters relate to. After each standard, the pertinent chapters are listed.

National Science Content Standards:
Structure and Function of Living Systems, Regulation and Behavior, and Scientific Inquiry

1. All organisms are composed of cells – the fundamental unit of life. Most organisms are single cells; other organisms, including humans, are multicellular. (3, 4)

2. Evidence for one-celled forms of life – the bacteria – extends back more than 3.5 billion years. (5)

3. Specialized cells perform specialized functions in multicellular organisms. (1, 2)

4. Groups of specialized cells cooperate to form a tissue, such as a muscle. Different tissues are in turn grouped together to form larger functional units, called organs. (2)

5. Each type of cell, tissue, and organ has a distinct structure and set of functions that serve the organism as a whole. (2)

6. Current scientific knowledge and understanding guide scientific investigations. (7)

7. Scientific explanations emphasize evidence, have logically consistent arguments, and use scientific principles, models, and theories. (7)

8. All organisms must be able to obtain and use resources, grow, reproduce, and maintain stable internal conditions while living in a constantly changing external environment. (2, 5, 6)

National Academies Press, 2005
http://www.nap.edu/readingroom/books/nses/

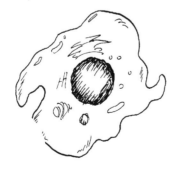

Overview of Source Cartoons

Consecutive rows that are similarly shaded indicate a set of cartoons that work together, and convey a plot when issued sequentially. The difficulty level ranges from easy (L1) to very challenging (L3).

Cartoon	Central Concept	Challenge Level	Helpful prior knowledge
Birds on the Run	5 Body Cells and their functions	L2	Proteins (Hemoglobin)
If you were a Cell	Unicellular Organisms vs. Body Cells	L1	Microorganisms
The Assistant	The Six Kingdoms	L2	Unicellular and Multicellular
The Six Kingdoms	The Six Kingdoms	L2	Archaebacteria
Cell Size	Role of the microscope in the discovery of cells	L1	History of Science, Metric Units
Two Types of Cells	Differences between prokaryotes and eukaryotes	L2	The Six Kingdoms
Microbes Rising	Microbes in the context of earth's history	L1	Geologic Time
Heated Microbes	Prokaryotes' ability to tolerate high temperatures	L1	Homeostasis
Hooked on Cells	Robert Hooke, who discovered cells	L1	Cell Walls
A Small World	How Microbes were discovered	L1	Microbiology, Early Microscopes
Schleiden and the Plants	Cell Theory	L2	Plant Cell Types
Meeting of the Minds	Cell Theory	L2	Plant and Animal Cells
Constructing a Theory	Cell Theory	L2	Hierarchy of Life
Rudolph Virchow	Cell Theory	L1	Cell Division

©Incentive Publications, Inc., Nashville, TN

Dr. Birdley Teaches Science – Classifying Cells

Teacher's Guide

<u>Contents</u>

Introduction   6

The Source Cartoon   7

Cartoon Profile   8

Assignments and Assessments   9

Preparing a Lesson   10

Sample Lesson Plan   11

A Hands-on Activity   12

The Source Cartoon

The Source Cartoon

The *Source Cartoon* explains the central concepts of the overall chapter. It is usually one or two pages in length. Expect to find the following in a given source cartoon:

• A central idea with supporting details

• Visual images related to the topic being presented

• Explanations of science concepts

• A range of characters that explain the information to each other or to the reader

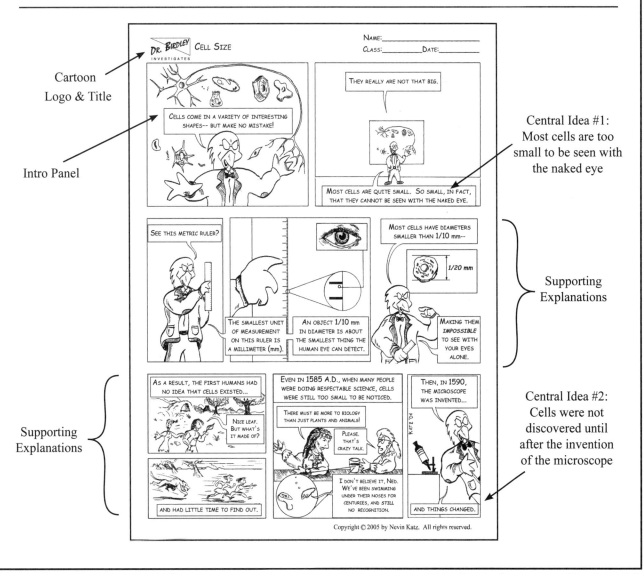

The Cartoon Profile

The *Cartoon Profile,* which outlines a source cartoon's science content, is useful for planning or teaching a lesson. Central elements include:

• The cartoon's objectives, vocabulary, main ideas, and related national standards.

• The "questions for discussion" below the image, which are useful for getting students engaged and checking for understanding.

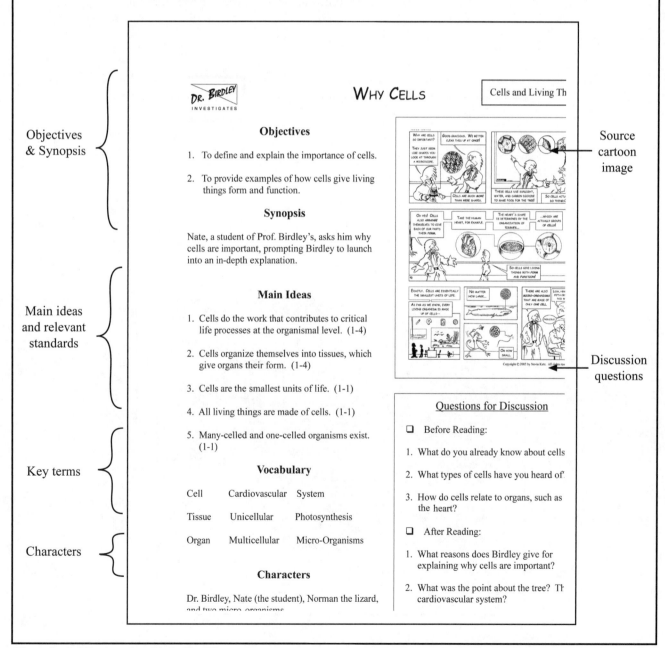

Assignments & Assessments

While a chapter's assignments help students build comprehension, quizzes can assess students' knowledge of key points from a given chapter. Five of the major assignments in this book and one quiz are pictured below.

Study Questions

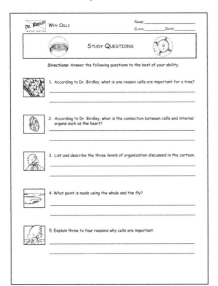

Visual Exercises

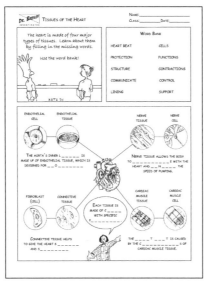

Graphic Organizer

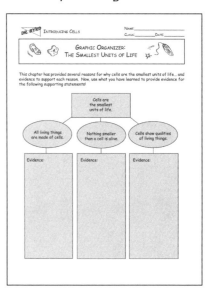

Vocabulary Build-up

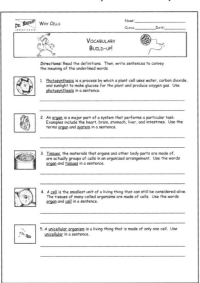

Background Article

Quiz

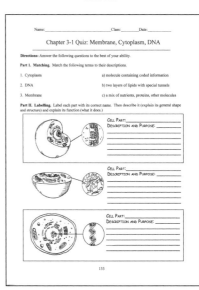

Sample Lesson Plan

This sample lesson plan integrates materials as a means of introducing cells. The overall format shown below can be applied to a range of different lessons.

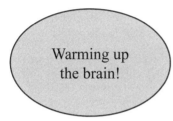

Lesson Objective: To define cells and explain their significance.

A. *Warm-up:* Examine cell under microscope in small groups. Report on what they see. Students list what they already know and questions they have about cells.

B. *Sharing ideas:* The teacher reviews the warm-up with students to learn about their prior knowledge.

C. *Vocabulary:* Students complete the vocabulary build-up, using key words in sentences.

A. *Introducing the Cartoon:* The teacher leads a discussion on the *Before Reading* questions from the cartoon profile.

B. *Classwide reading:* Several student volunteers read the cartoon aloud.

C. *Discussion:* The teacher leads a discussion on the *After Reading* questions from the cartoon profile.

D. *Reading in pairs:* Students read again in pairs, highlighting key words and writing comments on a separate sheet of paper.

A. *Independent Practice:* Students complete supplementary assignments, which include:
• study questions
• visual exercises
• background section questions
• graphic organizer
• Unit 1 quiz

Periodically, the class reviews the answers to the exercises.

B. *Activity:* Students examine more cell types under a microscope using different magnifications. They then draw each specimen.

A Hands-on Activity

Unit 1: Body Cells

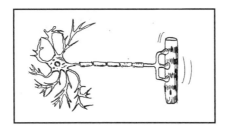

Contents

Source Cartoon: Birds on the Run   16

Cartoon Profile   17

Background   18

Study Questions   19

Mini-Comic: T-Cells   20

Visual Exercises (2)   21

Vocabulary Build-up   23

Panel Review   24

Graphic Organizer   25

Quiz   26

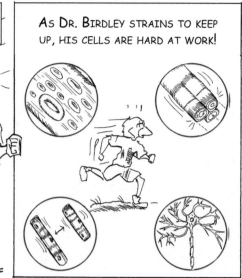
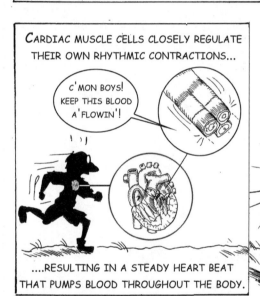
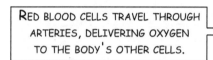
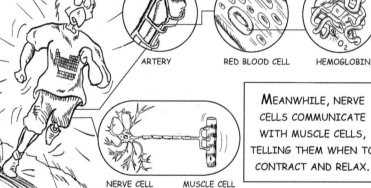
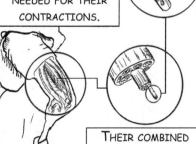

Birds on the Run

Cells and Body Systems

Objective

To show how four types of cells demonstrate key characteristics of life and help Dr. Birdley survive a long distance run.

Synopsis

As Dr. Birdley runs, his cells perform specific functions and show characteristics of life.

Main Ideas

1. Cardiac muscle cells communicate with each other to keep the heart beating.
2. To grab and release oxygen, red blood cells use hemoglobin, a molecule with a highly complex and orderly structure.
3. Nerve cells communicate to muscle cells in order to trigger contraction and relaxation.
4. Muscle cells use energy to do mechanical work.

Vocabulary

cardiac muscle cell	muscle cell
red blood cell	hemoglobin
nerve cell	contraction

Characters

Dr. Owen Birdley and his Dad

Teacher's Note

Despite the species difference, consider Birdley's anatomy to be highly similar to human anatomy. The mini-comic on T-Cells will be important for the second visual exercise and the quiz.

Questions for Discussion

Before Reading:

1. Think of a given body cell. What role does it play in your body?
2. What goes on inside your body during running or other sports?
3. How do cells help you in sports?

After Reading:

1. Who is in better shape: Owen or his Dad?
2. What is the role of a (given cell)?
3. What characteristics of life does each type of cell demonstrate?

DR. BIRDLEY INVESTIGATES — BODY CELLS

NAME:_____
CLASS:_____ DATE:_____

Background: Birds on the Run

The aim of this cartoon is twofold: to show the specific functions of four cell types, and to show how they demonstrate the characteristics of life. As Dr. Birdley (Owen) struggles through a seven-mile run, four body cells take center stage...

1. **Cardiac Muscle Cells.** These cells contract and relax in a rhythmic pattern in order to keep the heart pumping. They keep each other in sync by sending electrical signals to each other. By communicating in this way, these cells show behavior.

2. **Red Blood Cells.** Red blood cells deliver oxygen to other body cell parts. The ability of a red blood cell to grab, hold, and release oxygen is due to the highly orderly structure of a protein known as hemoglobin, which binds to oxygen molecules.

3. **Nerve Cells.** Nerve cells near the legs receive signals from the brain and transmit them to leg muscles using chemical and electrical signals. This essential form of communication is a good example of behavior.

4. **Muscle Cells.** During the long distance run, muscle cells use nutrients to produce useful energy, in the form of molecules known as ATP. This is an example of energy transfer. The rhythmic contractions can be thought of as behavior.

As you can see, Birdley's cells must cooperate intensely to help him get through this workout! Each cell has a specific job, and usually shows one or more characteristics of living things.

A final point is that the function of a particular cell contributes to the overall function of its tissue, organ, and system. For example, a muscle cell contracting relates directly to the contraction of a leg muscle. As a result, virtually any bodily function can be explained by how a group of cells are working together.

_____

For two different cells, explain a) the cell's specific job in the body, and b) how we know the cell is alive.

_____
_____
_____
_____

©Incentive Publications, Inc., Nashville, TN

Dr. Birdley Teaches Science – Classifying Cells

BIRDS ON THE RUN

NAME:_____
CLASS:_____ DATE:_____

 ## STUDY QUESTIONS

Directions: Answer the following questions to the best of your ability.

 1. How do cardiac muscle cells keep the heart beating properly?
_____
_____

 2. What helps red blood cells circulate around the body?
_____
_____

 3. What is the critical protein molecule that is located on red blood cells? Why is it important?
_____
_____

 4. Compare and contrast nerve cells with muscle cells. How are they similar? How are they different?
_____
_____

 5. What are two resources muscle cells need to produce mechanical motion?
_____
_____

 BODY CELLS

NAME:_____
CLASS:_____ DATE:_____

 MINI-COMIC: THE T-CELL

Directions: Read the cartoon and the text. Then, answer the question that follows.

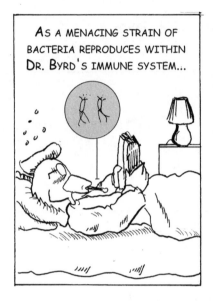
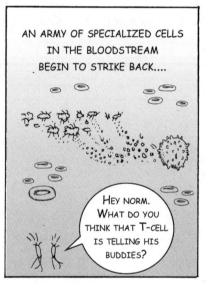
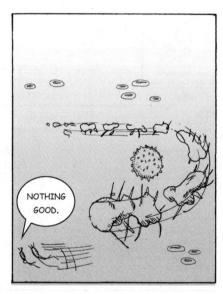

T-Cells, which travel throughout the body's bloodstream, are responsible for identifying foreign or "non-self" entities that do not belong in the body, such as viruses or harmful bacteria. T-cells are also known as white blood cells or T-lymphocytes. Some T-cells will strike back directly at the invader, while others communicate with other cells to mount a collective immune response. Here, a T-Cell uses chemical signals to tell a group of macrophages about two bacterial invaders. The macrophages, which used to be ordinary T-Cells, now specialize in hunting down and devouring "non-self" intruders.

Compare and contrast T-Cells with a) nerve cells or b) red blood cells.

_____

_____

_____

_____

©Incentive Publications, Inc., Nashville, TN   Dr. Birdley Teaches Science – Classifying Cells

LIVING BODY CELLS

NAME:_____
CLASS:_____ DATE:_____

IDENTIFY THE CHARACTERISTIC OF LIFE DEMONSTRATED BY EACH SPECIMEN. DEFEND YOUR ANSWER IN 1-2 SENTENCES.

CHARACTERISTICS OF LIFE

REPRODUCTION – PRODUCING NEW OFFSPRING

ENERGY TRANSFER – OBTAINING AND USING ENERGY

BEHAVIOR – RESPONDING TO THE ENVIRONMENT

ORGANIZATION – HAVING AN ORDERLY ARRANGEMENT OF PARTS

REGULATION – CONTROLLING PROPERTIES OR ACTIONS

GROWTH – DEVELOPMENT OVER TIME

1. THE COMPLEX HEMOGLOBIN MOLECULES SIT ATOP THE RED BLOOD CELL. THE HIGHLY ORDERLY STRUCTURE OF HEMOGLOBIN ALLOWS IT TO GRAB ONTO OXYGEN MOLECULES.

_____
_____
_____
_____

2. MUSCLE CELLS USE OXYGEN AND NUTRIENTS TO PRODUCE USEFUL ENERGY. THE MITOCHONDRION WITHIN THE CELL MAKES THIS HAPPEN.

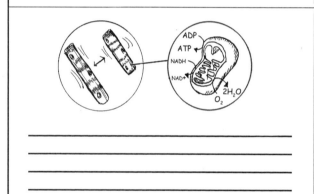

_____
_____
_____
_____

3. NEURONS SEND SIGNALS TO STRIATED MUSCLE CELLS, WHICH RESPOND BY CONTRACTING AND RELAXING.

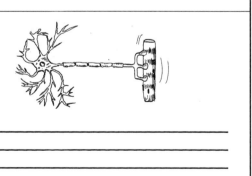

_____
_____
_____
_____

4. CARDIAC MUSCLE CELLS COMMUNICATE TO EACH OTHER USING ELECTRICAL IMPULSES, ENABLING THEM TO ALL MAINTAIN THE SAME RHYTHM.

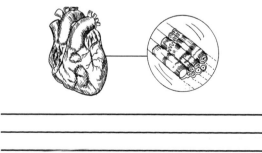

_____
_____
_____
_____

Body Cell Types

NAME:_____
CLASS:_____ DATE:_____

For each cell type, explain where it is in the body and what it does. Be as detailed as possible!

Locations	Functions
Bloodstream (2)	Mechanical Motion
Leg Muscle	Sensing
Heart	Controlling the Body
Throughout Body	Delivering Oxygen
Brain	Creating Heart Beat
Bicep Muscle	Fighting and Preventing Illness

1. Cardiac Muscle Cells _____

2. Red Blood Cells _____

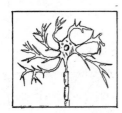

3. Nerve Cells_____

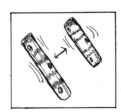

4. Striated Muscle Cells_____

5. T-Cells _____

©Incentive Publications, Inc., Nashville, TN

Dr. Birdley Teaches Science – Classifying Cells

 BIRDS ON THE RUN

NAME:_____
CLASS:_____DATE:_____

VOCABULARY BUILD-UP!

Directions: Read the definitions. Then, write sentences to convey the meaning of the underlined words.

1. Cardiac muscle cells are cells in the heart that tighten and loosen (contract and relax) together so the heart beats properly. Use cardiac muscle cells in a sentence:

_____

_____

2. Red blood cells are cells in the bloodstream that pick up oxygen at the lungs, and drop it off at other body parts. They do not move by themselves, but are pumped through the body by the heart. Use red blood cells in a sentence:

_____

_____

3. Hemoglobin is a protein molecule on the red blood cell that "grabs" oxygen while it is in the lungs, and then transports it to the other cells in the body. Use hemoglobin in a sentence:

_____

_____

4. Nerve cells specialize in communicating. Some of them transmit signals from the brain to the muscles and back. Use nerve cells in a sentence:

_____

_____

5. Striated muscle cells are muscle cells that YOU can control. Your legs are made up of striated muscle fibers. When they contract and relax, your legs can move. They need oxygen and nutrients to produce useful energy. Use striated muscle cells in a sentence:

_____

_____

©Incentive Publications, Inc., Nashville, TN

Dr. Birdley Teaches Science – Classifying Cells

NAME:_____
CLASS:_____ DATE:_____

PANEL REVIEW: CONNECTING TERMS

Directions: Review the panels in the space below and answer the questions that follow.

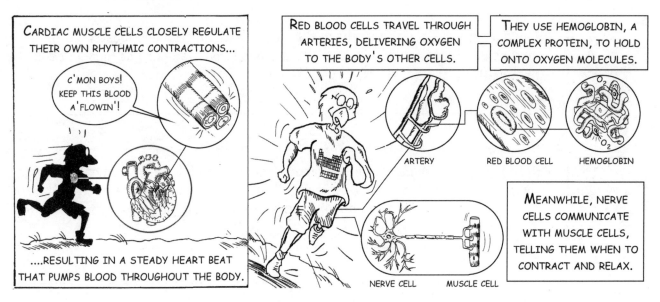

1. Explain how the heart relates to cardiac muscle cells.

2. Explain how blood cells relate to the heart.

3. Explain how blood cells relate to muscle cells.

4. Explain how nerve cells relate to muscle cells.

DR. BIRDLEY INVESTIGATES — BIRDS ON THE RUN

NAME:_____
CLASS:_____ DATE:_____

 # Graphic Organizer

Directions: List five cell types as well as their locations and functions.

Cell Type and Location:	Function:
Cell Type and Location:	Function:
Cell Type and Location:	Function:
Cell Type and Location:	Function:
Cell Type and Location:	Function:

©Incentive Publications, Inc., Nashville, TN

Dr. Birdley Teaches Science – Classifying Cells

Name:_____ Class:_____ Date:_____

Unit 1 Quiz: Body Cells

Directions: This quiz tests your knowledge of the chapter's cartoon, background article, and visual exercises. Answer the following questions to the best of your ability.

1. The primary function of nerve cells is to:
   - (a) recognize and destroy foreign substances
   - (b) direct movement, thinking, and learning
   - (c) deliver oxygen to other body cells
   - (d) contract and relax in order to do mechanical work

2. The primary function of cardiac muscle cells is to:
   - (a) generate the heart beat
   - (b) contract and relax all at the same time
   - (c) deliver oxygen to body cells
   - (d) open and close the valves of the heart

3. Which of the following best describes the function of hemoglobin in red blood cells?
   - (a) provides nutrients to neighboring tissues
   - (b) direct movement, thinking, and learning
   - (c) deliver oxygen to other body cells
   - (d) contract and relax in order to do mechanical work

4. If someone gets sick, which of the following cells is most likely to respond directly to the bacterial infection?
   - (a) T-cells
   - (b) nerve cells
   - (c) striated muscle cells
   - (d) red blood cells

5. The use of energy to generate voluntary mechanical motion is most directly accomplished by:
   - (a) T-cells
   - (b) nerve cells
   - (c) striated muscle cells
   - (d) red blood cells

6. Red blood cells and nerve cells can both be found within the same
   - (a) tissue
   - (b) organelle
   - (c) system
   - (d) organism

7. Compare and contrast the functions of two types of cells.

_____

_____

_____

_____

©Incentive Publications, Inc., Nashville, TN

Dr. Birdley Teaches Science – Classifying Cells

Unit 2: From a Cell's Point of View

Contents

Source Cartoon: If You Were a Cell   26

Cartoon Profile   27

Background   28

Study Questions   29

Visual Exercise   30

Vocabulary Build-up   31

Sample Homework Assignment   32

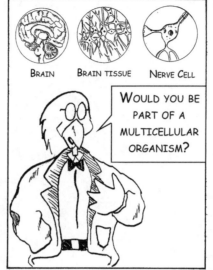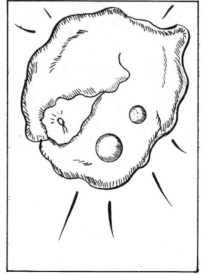

If You Were a Cell

Types of Cells

Objective

To provoke thought towards writing a story from the perspective of a cell.

Synopsis

Dr. Birdley challenges his students to write the story from the perspective of a cell.

One student persuades the reader to write as a body cell from a multicellular organism. Another student argues that unicellular organisms have better lives. A microbe that boasts of its independence gets captured by an amoeba.

Main Ideas

1. Body cells, which are parts of multicellular organisms, must work as a team to keep the organism alive.

2. Unicellular organisms have more independence, but live in danger of being eaten by larger one-celled organisms.

Vocabulary

nerve cell   unicellular organism
brain   multicellular organisms
muscle fiber   bacterium

Characters

Dr. Birdley, Students (Celia and Dan), Bacterium, Amoeba, Norman the Lizard

Teacher's Notes

Use the homework sample and "cell story starters" exercise as resources for helping students write their own cell stories.

Questions for Discussion

Before Reading:

1. If you could be any type of cell, what cell type would you be? Why?

2. Would you want to be a body cell or a one-celled organism? Why or why not?

After Reading:

1. Why does Norman the lizard seem to prefer the "multicellular" deal?

2. What are the advantages and disadvantages of being a body cell versus a one-celled organism?

3. Did the argument(s) in the cartoon change your initial opinion? Why or why not?

©Incentive Publications, Inc., Nashville, TN

Dr. Birdley Teaches Science – Classifying Cells

FROM A CELL'S POINT OF VIEW

NAME:_____
CLASS:_____ DATE:_____

 ## Background: If You Were a Cell

This cartoon has two objectives. The first is to raise the question: if you were a cell, what type of cell would you be? The second goal is to distinguish between multicellular and unicellular organisms.

In choosing a cell to write about, consder the following two options: you can choose between a unicellular organism or a body cell in a multicellular organism.

Two examples of body cells are illustrated. The first example is a nerve cell that is part of the brain tissue. The second example is a striated muscle fiber, a cell within the bicep muscle tissue.

Nerve cells in the brain constantly communicate, sending chemical and electrical signals, so that the brain can effectively learn and control the body. When learning takes place, nerve cells build more connections with each other, called synapses.

In the bicep, cells must contract and relax at the proper times so that the arm can move properly. They also must respond to nerve cells, who are sending them signals from the brain.

Being a body cell involves a lot of teamwork, but your life is devoted to serving the whole organism. You can't come and go as you please... you have to stay at your post and perform your function. Benefits include an environment protected by the immune system, as well as a steady supply of nutrients.

If you are a unicellular organism, you have more independence because you are not part of a larger system. The downside is that there is potentially less safety. The environment is not the inside of a body...it is an ecosystem filled with other microorganisms — which are sometimes bigger than you! As a result, you have to find food and escape from predators on your own.

_____

1. Explain one advantage and one disadvantage of being a body cell in a multicellular organism.

_____

_____

2. Explain one advantage and one disadvantage of being a unicellular organism.

_____

_____

IF YOU WERE A CELL

NAME:_____
CLASS:_____ DATE:_____

  STUDY QUESTIONS

Directions: Answer the following questions to the best of your ability.

   1. If you could be any type of cell, what cell would you be? Why?
_____
_____

   2. How do nerve cells relate to the brain?
_____
_____

   3. What is one advantage and one disadvantage of being a unicellular (single-celled) organism?
_____
_____

   4. What is one advantage and one disadvantage of being a cell in a multicellular organism?
_____
_____

   5. What environment would you live in as a unicellular organism?
_____
_____

©Incentive Publications, Inc., Nashville, TN

Dr. Birdley Teaches Science – Classifying Cells

Cell Story Starters

Name: _____
Class: _____ Date: _____

In each panel, fill in the blanks with the words in the bank. Use these panels as inspiration for writing your own story!

Word Bank

BICEP	COMMUNICATE
MOVE	LEARNING
FIBERS	SYNAPSES
ENERGY	PROTISTS
CONTRACT	BACTERIUM
EXCESS	CILIA

---

1. I am a muscle fiber. I work with other _ _ _ _ _ _ s to help the _ _ C _ _ _ muscle to _ _ V _.

BICEP   MUSCLE FIBER BUNDLE   MUSCLE FIBER (CELL)

We receive signals from nerve cells on when to _ _ _ _ _ _ _ T or relax. Our job takes a lot of _ _ _ _ _ Y.

---

2. I am a neuron. I c _ _ _ _ _ _ _ _ _ _ _ with other cells to help the organism think, sense, move, and learn.

BRAIN   BRAIN TISSUE   NERVE CELL

I build connections with other nerve cells, called _ _ N _ _ _ _ S, in order to make _ _ _ _ N _ _ G happen.

---

3. I am a paramecium. I use tiny hairs called C _ _ _ _ to move around and chase down other bacteria!

I contain everything I need to survive on my own. Food enters me through a pore lined with cilia. I also have star-shaped **contractile vacuoles** which help me to pump out E _ _ _ _ _ water.

---

4. I am a bacterium. I am made of one cell. I enjoy living in hot places, where P _ _ _ _ _ _ _ _ can't find me!

My hobbies include dividing in two and exchanging genetic material, or **DNA**, with other B _ _ _ _ _ _ _ _.

IF YOU WERE A CELL

NAME:_____
CLASS:_____ DATE:_____

 VOCABULARY BUILD-UP!

Directions: Read the definitions. Then, write sentences to convey the meaning of the underlined words.

1. The <u>biceps brachii</u> is a muscle in our upper arm that enables us to rotate or move our forearms. Use <u>biceps brachii</u> in a sentence.

_____

_____

2. A <u>muscle fiber bundle</u> consists of many muscle fibers working together in order to generate movement. Use <u>muscle fiber bundle</u> in a sentence.

_____

_____

3. A <u>muscle fiber</u> is a muscle cell that performs mechanical work by contracting and relaxing. Use <u>muscle fiber</u> in a sentence.

_____

_____

4. <u>E. Coli</u> is type of single-celled bacterium. Use the term <u>E. Coli</u> in a sentence.

_____

_____

5. <u>Phagocytosis</u> is the process by which some cells engulf their prey in order to obtain nutrients and energy. Use <u>phagocytosis</u> in a sentence.

_____

_____

©Incentive Publications, Inc., Nashville, TN   Dr. Birdley Teaches Science – Classifying Cells

If You Were a Cell

NAME:_____
CLASS:_____ DATE:_____

 # Sample Homework Assignment

>   Nate McCoy
>   Dr. Birdley
>   5/25/10
>
> *If You Were a Cell: Nerve Cells in the Brain*
>
> I am a nerve cell, otherwise known as a neuron, in the brain of a human named Bob. I have built connections with other neurons so that I can communicate with them. I have a cool shape with all these branches. A long branch, called the axon, allows me to send signals. Small branches, called the dendrites, allow me to receive signals.
>
> Each part of the brain contains a different network that performs a specific function. In my region, the cerebral cortex, we produce thought, speech, and voluntary movement. If Bob is learning something, I may be building connections with other neurons. Other nerve cells regulate breathing, balance, and emotions. Still others are also responsible for receiving and sorting the information from Bob's five senses. That is all for now—I must get back to work.

Study Questions

1. What is the function of the nerve cell?

2. Where are nerve cells located?

3. What does the nerve cell use to send signals? receive signals?

4. What region of the brain is discussed? What does this region do?

Unit 3: Cells and the Kingdoms

Contents

Source Cartoon: The Assistant   34

Source Cartoon: Kingdoms Overview   35

Cartoon Profiles (2)   36 and 37

Background   38

Study Questions   39

Visual Exercise   40 and 41

Vocabulary Build-up   42

Graphic Organizer   43

Unit Quiz   44

The Assistant

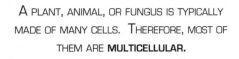

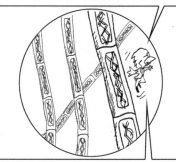

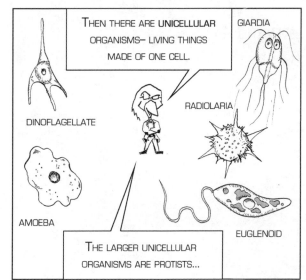

Kingdoms Overview

Name:_____
Class:_____ Date:_____

Animalia

Multicellular organisms that consist of vertebrates and invertebrates.

Plantae

This kingdom includes multicellular organisms that use **photosynthesis!** Some of us eat, though I'd say it's more of a hobby.

Fungi

Organisms that break down organic matter by secreting chemicals!

Most of them, like mushrooms, are multicellular...

...but some, like yeast, are unicellullar.

Protista

A protist may be a one-celled living thing...

...or simple many-celled organism.

Eubacteria

Unicellular organisms that do not have nuclei. They are found in your gut, in ponds, on doorknobs, and many other places.

Like protists, they were so small that they were not discovered until after the invention of the microscope.

Archaebacteria

Evidence suggests that archaebacteria were the first organisms on Earth.

Many of them prefer environments that are extremely hot, unusually cold, highly salty, or devoid of oxygen!

©Incentive Publications, Inc., Nashville, TN
©Nevin Katz

Dr. Birdley Teaches Science – Classifying Cells

THE ASSISTANT

Cells and Taxonomy

Objectives

1. To establish which types of organisms are multicellular and unicellular.
2. To introduce how organisms are categorized into kingdoms.

Synopsis

Jaykes' new computer animation, Digital Birdley, explains which types of organisms are multicellular, unicellular, or both.

Main Ideas

1. Animals, plants, and fungi are typically multicellular. One exception is yeast, which is a unicellular fungus.
2. Some protists are multicellular, but not as complex as plants, animals, or fungi.
3. Other protists are unicellular. They are larger than bacteria.
4. Bacteria are virtually all unicellular, and are smaller than protists.

Vocabulary

animals	protists	multicellular
plants	bacteria	unicellular
fungi	spirogyra	euglenoid

Characters

Dr. Birdley, Jaykes (the programmer), and the new assistant, Digital Birdley

Teacher's Note

Use the "Kingdoms Overview" Cartoon in this chapter as an additional resource.

Questions for Discussion

Before Reading:

1. How would you classify the major types of living things?
2. How have scientists used cells to help classify living things?

After Reading:

1. Which type of organism do you find most interesting in the cartoon?
2. What differences do you see among the one-celled organisms?
3. How is the spirogyra different from the other protists in the cartoon?

Kingdoms Overview

Cells and Taxonomy

Objectives

To introduce the six kingdoms and point out the main types of organisms categorized under each kingdom.

Synopsis

Each panel contains a quick description of a particular kingdom, along with an example.

Main Ideas

1. Animalia include multicellular organisms that are both invertebrates and vertebrates.

2. Plantae includes multicellular organisms that use photosynthesis.

3. Fungi break down matter to extract energy.

4. Protists have nuclei in their cells and include unicellular and simple multicellular organisms.

5. Eubacteria include one-celled organisms that are small and do not have a nucleus.

6. Many archaebacteria prefer environments that other living things would find intolerable.

Vocabulary

animalia   fungi   eubacteria
plantae   protista   archaebacteria

Characters

Dr. Birdley and additional organisms

Teacher's Note

This cartoon works well as a transparency. It is designed as additional support for the study questions and visual exercise in this chapter.

Questions for Discussion

Before Reading:

1. What are the major types of living things that exist?

2. Which type of living things are made of one cell? Which are made of many cells?

After Reading:

1. Which kingdoms include unicellular organisms?

2. How are plants different from fungi in their methods of obtaining energy?

3. What are some common traits of archaebacteria?

©Incentive Publications, Inc., Nashville, TN

Dr. Birdley Teaches Science – Classifying Cells

THE SIX KINGDOMS

NAME:_____
CLASS:_____ DATE:_____

Background: The Assistant Kingdoms' Overview

The central purpose of this cartoon is to point out which types of organisms are multicellular, unicellular, or both. Read on as each kingdom will be discussed.

Animalia. Animals are virtually all multicellular, or made up of many cells. Many animals have the four levels of organization known as systems, organs, tissues, and cells.

Plantae. Plants are multicellular. Plant cells are different from animal cells in that they are able to conduct photosynthesis and generally have more space for holding water inside them. They also have sturdy cell walls surrounding them.

Fungi. Fungi are also multicellular. Many produce offspring by making reproductive cells known as spores. Unlike plants, fungi do not make their own food, but survive by absorbing nutrients from organic materials. They send out chemicals that break down the material before absorbing it.

Protista. Protists, small microorganisms, are either multicellular or unicellular. The amoeba is an example of a unicellular protist. The spirogyra in the central picture is a multicellular protist.

Most of the multicellular protists are plant-like, while the animal-like protists, such as the amoeba, are virtually always unicellular.

Bacteria previously formed a single kingdom known as monera, but it has recently been separated into two kingdoms: eubacteria and archaebacteria.

Eubacteria. Bacteria are always unicellular, and smaller than the protists. Some form groups known as colonies.

Archaebacteria. These bacteria have radically different DNA from eubacteria. Many live in extremely hot environments...sometimes with temperatures above boiling. They are also unicellular.

---

1. List two types of multicellular organisms and explain how they are different.

_____

_____

2. List two types of unicellular organisms and explain how they are different.

_____

_____

THE ASSISTANT & KINGDOMS OVERVIEW

NAME:_____
CLASS:_____ DATE:_____

  STUDY QUESTIONS

Directions: Answer the following questions to the best of your ability.

1. What is the difference between multicellular and unicellular organisms?
_____
_____

2. Which three kingdoms contain living things that are virtually always multicellular? Give one example of a living thing from each kingdom.
_____
_____

3. Which kingdoms contain both unicellular and multicellular living things? Give two examples of living organisms from each kingdom.
_____
_____

4. Although spirogyra uses photosynthesis, it is different from the plants and trees we see every day. How do you think it might be different?
_____
_____

5. Earlier in history, scientists used fewer kingdoms for grouping living things. At one point, they used only two groups: plants and animals. Why do you think this was the case?
_____
_____

©Incentive Publications, Inc., Nashville, TN

Dr. Birdley Teaches Science – Classifying Cells

 CLASSIFY IT!

NAME:_____
CLASS:_____ DATE:_____

A. IDENTIFY EACH ORGANISM AS UNICELLULAR OR MULTICELLULAR.

B. IDENTIFY THE KINGDOM IT BELONGS TO.

C. DEFEND YOUR ANSWER!

TERMS
Multicellular   Unicellular

KINGDOMS
Animalia   Plantae

Fungi   Protista

Eubacteria   Archaebacteria

---

1. Mushrooms obtain nutrients by secreting chemicals that break down organic materials.

_____
_____
_____
_____

2. The euglenoid not only eats food, but makes it using photosynthesis. Notice its large size.

_____
_____
_____
_____

3. These microbes inhabit environments with moderate temperatures.

Micrococcus Cryophilus   Bacillus Psychrophilus

_____
_____
_____
_____

4. These guys need water, sunlight, carbon dioxide, and nutrients to grow.

_____
_____
_____
_____

©Incentive Publications, Inc., Nashville, TN

Dr. Birdley Teaches Science – Classifying Cells

CLASSIFY IT!

Name:_____
Class:_____ Date:_____

5. Although more complex than many microbes, the spirogyra does not have organs or systems as we know them.

_____
_____
_____

6. The chameleon's cells change shape, causing its skin to change color.

_____
_____
_____

7. These microbes live in sulfur pools with temperatures above 100°C.

_____
_____
_____
_____

8. While the dinoflagellate makes its food using photosynthesis, the giardia eats to stay alive. Both move using tails called flagella.

Dinoflagellate   Giardia

_____
_____
_____
_____

9. The aardvark is the only species in the mammalian order **Tubulidentata**.

_____
_____
_____

10. Neither a bacterium nor a protist, this one-celled yeast reproduces by a process known as budding.

_____
_____
_____

©Incentive Publications, Inc., Nashville, TN

Dr. Birdley Teaches Science – Classifying Cells

THE ASSISTANT

NAME:_____
CLASS:_____ DATE:_____

 VOCABULARY BUILD-UP!

Directions: Read the definitions. Then, write sentences to convey the meaning of the underlined words.

1. A <u>multicellular</u> organism is a living thing made up of many cells. Use <u>multicellular</u> in a sentence.

   _____

   _____

2. A <u>unicellular</u> organism is a living thing made up of only one cell. Use <u>unicellular</u> in a meaningful sentence.

   _____

   _____

3. A <u>protist</u> is a tiny living thing that is either unicellular or multicellular. Use <u>protist</u> in a sentence.

   _____

   _____

4. <u>Spirogyra</u> is a plant-like protist that is multicellular. Use <u>spirogyra</u> in a sentence.

   _____

   _____

5. <u>Bacteria</u> are the smallest organisms on earth. They are typically unicellular. Use <u>bacteria</u> in a sentence.

   _____

   _____

©Incentive Publications, Inc., Nashville, TN

Dr. Birdley Teaches Science – Classifying Cells

INTRODUCING CELLS

NAME:_____
CLASS:_____ DATE:_____

Graphic Organizer: The Smallest Units of Life

This chapter has provided several reasons for why cells are the smallest units of life...and evidence to support each reason. Now, use what you have learned to provide evidence for the following supporting statements!

Cells are the smallest units of life.

- All living things are made of cells.
  - Evidence:

- Nothing smaller than a cell is alive.
  - Evidence:

- Cells show qualities of living things.
  - Evidence:

©Incentive Publications, Inc., Nashville, TN

Dr. Birdley Teaches Science – Classifying Cells

Name:_____ Class:_____ Date:_____

Unit 3 Quiz: Cells and the Kingdoms

Directions: Answer the following questions to the best of your ability.

1. Which of the following organisms belongs to the kingdom **eubacteria**?

   (a) yeast   (b) spirogyra

   (c) bacillus psychrophilus   (d) mushrooms

2. One example of a multicellular organism is the
   (a) spirogyra
   (b) amoeba
   (c) bacteria
   (d) T4 bacteriophage virus

3. The smallest unicellar organisms are classified as:
   (a) protists
   (b) bacteria
   (c) fungi
   (d) plantae

4. A scientist is studying two photosynthetic organisms under a microscope. One is multicellular while the other is unicellular. Both belong to the same kingdom. They are most likely classified as:
   (a) protists
   (b) bacteria
   (c) animals
   (d) plants

5. A multicellular organism obtains its energy primarily by secreting chemicals that break down dead matter. This organism is most likely to be classified under:
   (a) fungi
   (b) plantae
   (c) animalia
   (d) protista

6. Compare and contrast a cell from the leaf of a plant to the **euglenoid**, a unicellular photosynthetic protist. How are they similar? How are they different?

_____

_____

_____

_____

Unit 4: Two Types of Cells

Contents

Source Cartoon: Two Types of Cells   46

Cartoon Profile   47

Background   48

Study Questions   49

Visual Exercises (2)   50 and 51

Vocabulary Build-up   52

Concept Map   53

Unit Quiz   54

DR. BIRDLEY INVESTIGATES: TWO TYPES OF CELLS

NAME:_____
CLASS:_____ DATE:_____

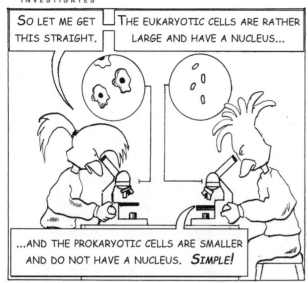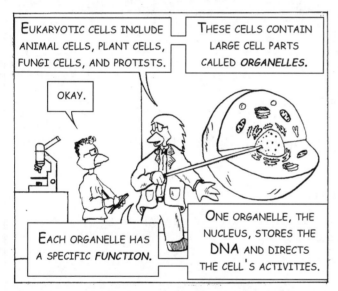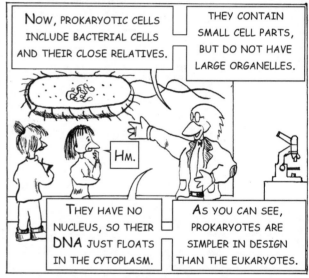

©Incentive Publications, Inc., Nashville, TN
©Nevin Katz

Dr. Birdley Teaches Science – Classifying Cells

Two Types of Cells

Classifying Cells

Objectives

To explain both the obvious and more technical differences between prokaryotic and eukaryotic cells.

Synopsis

After students distinguish between prokaryotic and eukaryotic cells through a microscope, Dr. Birdley teaches that the inside of the two cells are very different.

Main Ideas

1. Both prokaryotic and eukaryotic cells have sufficient complexity to allow them to carry out life functions.

2. The life functions within a eukaryotic cell are compartmentalized within large organelles.

3. The life functions of a bacterial cell occur within the cytoplasm.

4. All plants, animals, fungi, and protists contain eukaryotic cells.

5. Prokaryotic cells are mainly bacteria.

6. While eukaryotic DNA is in the nucleus, the DNA of prokaryotes floats in the cytoplasm.

Vocabulary

prokaryotes	nucleus	fungi
eukaryotes	animals	protists
organelles	plants	bacteria

Characters

Dr. Birdley, students, irate bacteria

Questions for Discussion

Before Reading:

1. What are some differences and similarities have you have noticed among the cells you have seen so far?

2. What differences exist between bacteria and other living things?

After Reading:

1. Why are the bacteria offended?

2. How are prokaryotes different from eukaryotes? How are they similar?

3. Are all microorganisms prokaryotic? If your answer is no, which ones are not?

©Incentive Publications, Inc., Nashville, TN

Dr. Birdley Teaches Science – Classifying Cells

PARTS OF THE CELL

NAME:_____

CLASS:_____ DATE:_____

 # Background: Two Types of Cells

The purpose of this cartoon is to differentiate between two major types of cells: prokaryotes and eukaryotes.

The cartoon opens up with two students introducing basic differences between the two: whereas prokaryotes are generally small cells with no nucleus, eukaryotes are larger and always contain a nucleus.

Dr. Birdley goes on to point out that animals, plants, fungi, and protists are all made up of eukaryotic cells. All these organisms, except for one-celled protists, are multicellular. Prokaryotes, on the other hand, include one-celled bacteria and their cousins.

A key difference between "proks" and "euks" is that eukaryotes are more complex, and have many of their functions compartmentalized into organelles. For example, the nucleus is responsible for controlling the cell, the lysozome breaks down waste, and the mitochondrion produces useful energy. In contrast, prokaryotic cells do not have large organelles, and all cellular activities occur in the cytoplasm.

Both eukaryotes and the organisms they are a part of are more complex than prokaryotic cells. Because all this complexity needs to be "programmed" in genetic instructions, eukaryotes have significantly more DNA than prokaryotes.

Both prokaryotes and eukaryotes share several attributes: they all contain a membrane, DNA, and cytoplasm. They are all the smallest units of life. Finally, most of them are too small to be seen with the naked eye.

___

1. Explain how prokaryotes are different from eukaryotes in terms of internal cell structure.
___
___

2. If you were looking at a cell under a microscope, how could you identify it as a prokaryote or eukaryote?
___
___

→ Note: Red blood cells are an exception. They are eukaryotes without a nucleus.

Two Types of Cells

Name:_____
Class:_____ Date:_____

Study Questions

Directions: Answer the following questions to the best of your ability.

1. What are the two major types of cells being discussed? What differences between them appear through the light microscope in the first panel?
_____
_____

2. Explain how the eukaryotic cell is internally different from the prokaryotic cell.
_____
_____

3. Explain how eukaryotes are different from prokaryotes in terms of their DNA.
_____
_____

4. What four kingdoms have organisms that contain eukaryotic cells?
_____
_____

5. What type of cell speaks to Dr. Birdley? Why was it offended? Explain why Birdley might have made the offensive comment.
_____
_____

©Incentive Publications, Inc., Nashville, TN

Dr. Birdley Teaches Science – Classifying Cells

DR. BIRDLEY INVESTIGATES: CELL TYPE IDENTIFICATION

NAME:_____
CLASS:_____ DATE:_____

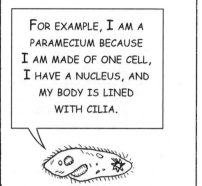

These look like _____
_____
_____
_____
_____

Possible Specimens include:
nerve cells   algae
plant cells   euglenoids
amoeba   smooth muscle cells

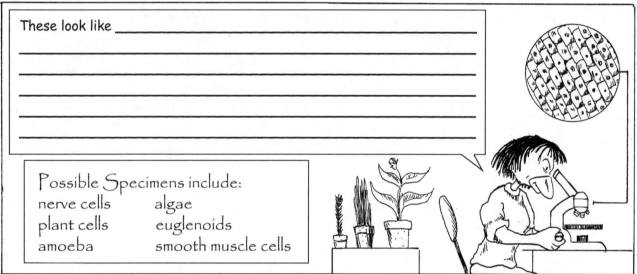

I think these are _____
_____
_____
_____
_____

Possible Specimens include:
plant cells   red blood cells
T-cells   bacteria
amoeba   smooth muscle cells

©Incentive Publications, Inc., Nashville, TN

Dr. Birdley Teaches Science – Classifying Cells

DR. BIRDLEY INVESTIGATES

CELL TYPE IDENTIFICATION, CONTINUED

NAME: ______________________

CLASS: __________ DATE: __________

These are ____

____

____

____

____

____

Possible cell types include:

amoeba	onion cells
nerve cells	euglenoids
plant cells	bacteria

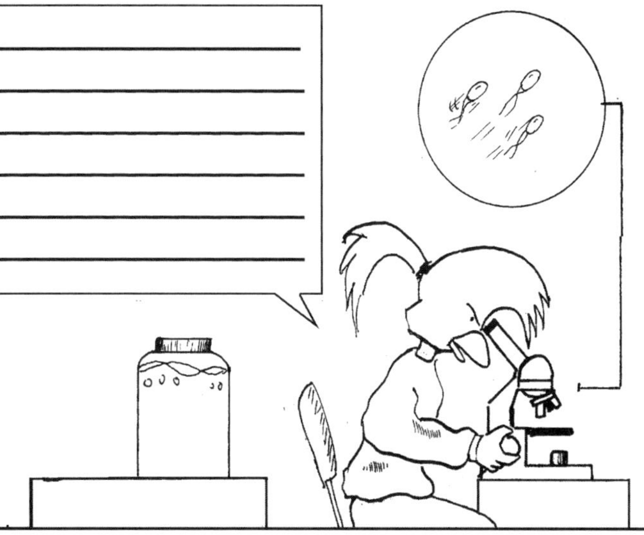

These look like ____

____

____

____

____

____

Possible cell types include:

amoeba	onion cells
nerve cells	euglenoids
plant cells	smooth muscle cells

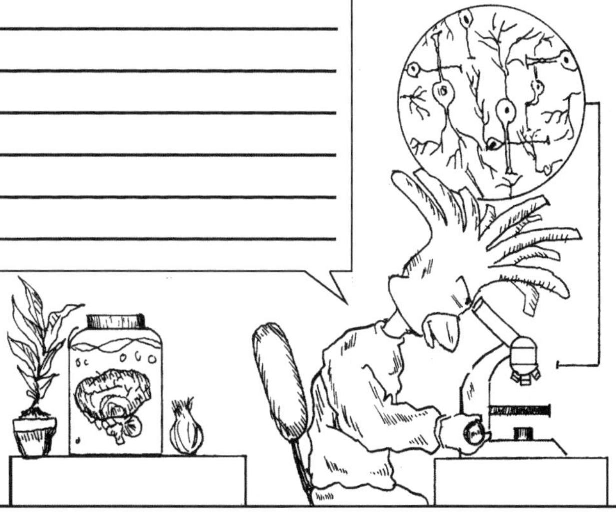

I have identified these as ____

____

____

____

____

____

Possible cell types include:

amoeba	euglenoids
bacteria	cardiac muscle cells
plant cells	smooth muscle cells

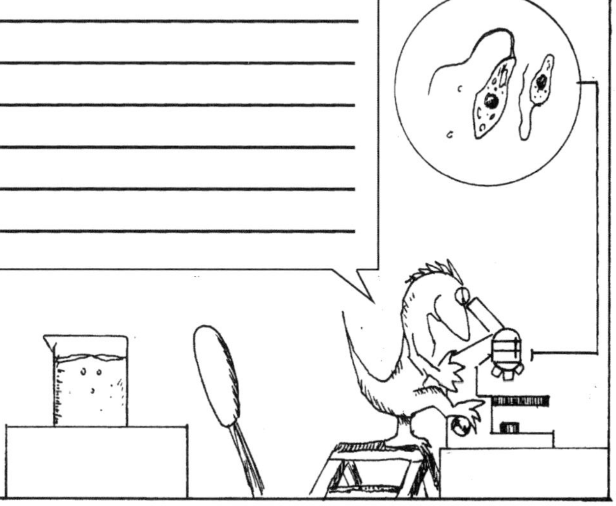

Two Types of Cells

NAME:_____
CLASS:_____ DATE:_____

 **Vocabulary Build-up!**

Directions: Read the definitions. Then, write sentences to convey the meaning of the underlined words.

1. <u>Prokaryotes</u>, which include bacterial cells and their cousins, are small one-celled organisms that are simple in structure, but genetically advanced (They have had 3.5 billion years to evolve.) Use <u>prokaryote</u> in a sentence.
_____
_____

2. <u>Eukaryotes</u> are complex cells that contain larger, membrane-bound organelles. Animals, plants, fungi, and protists, are composed of eukaryotic cells. Use <u>eukaryote</u> in a sentence.
_____
_____

3. A <u>membrane-bound organelle</u> is a cell part that is enclosed or bound by its own membrane. Like all organelles, these cell parts have specific functions. Use the term <u>membrane-bound organelle</u> in a sentence.
_____
_____

4. The <u>nucleus</u> is a membrane-bound organelle within eukaryotes that stores DNA and controls the cell's activities. Use the term <u>nucleus</u> in a sentence.
_____
_____

5. The <u>nucleoid</u> is the cluster of DNA strands within prokaryotic cells that serves as the primary source of genetic information. It is not enclosed by a nucleus, but floats freely in the cytoplasm. Use <u>nucleoid</u> in a sentence.
_____
_____

©Incentive Publications, Inc., Nashville, TN

Dr. Birdley Teaches Science – Classifying Cells

DR. BIRDLEY INVESTIGATES — TWO TYPES OF CELLS

NAME:_____
CLASS:_____ DATE:_____

Concept Map

Fill in the bubbles with the words from the word bank. Use the source cartoon "Two Types of Cells" and the related background section to help.

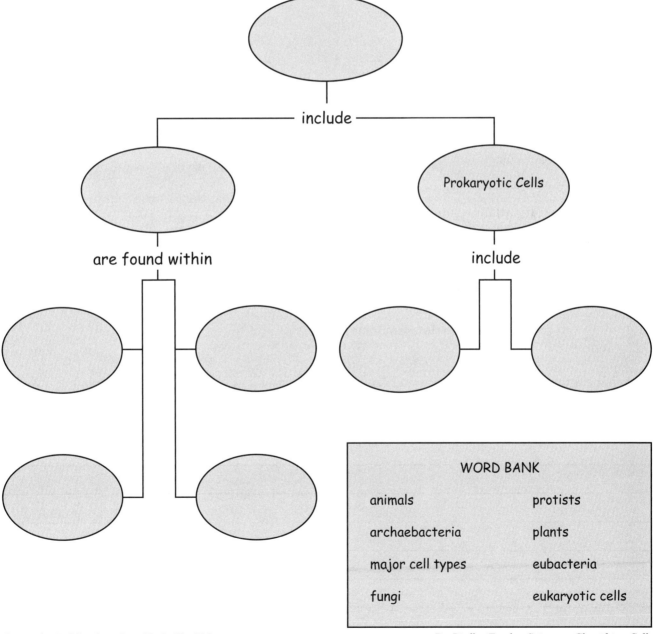

WORD BANK

animals   protists

archaebacteria plants

major cell types eubacteria

fungi   eukaryotic cells

Dr. Birdley Teaches Science – Classifying Cells

Name:_____ Class:_____ Date:_____

Unit 4 Quiz: Two Types of Cells

Directions: This quiz tests your knowledge of the chapter's cartoon, background article, and visual exercises. Answer the following questions to the best of your ability.

1. Which of the following is NOT a characteristic of a eukaryotic cell?
   a) DNA floats freely in the cytoplasm
   b) contains a nucleus
   c) typically larger than bacteria
   d) large organelles

2. Which of the following organisms is prokaryotic?
   a) mushroom
   b) amoeba
   c) bacterium
   d) millipede

3. Organelles are all similar in that they all:
   a) store DNA
   b) can be found in prokaryotic cells
   c) perform specific functions within the cell
   d) are located within the membrane

4. A plant cell is
   a) prokaryotic
   b) eukaryotic

5. Cells where the functions are more compartmentalized are
   a) prokaryotic
   b) eukaryotic

6. Cells that are relatively simple in structure are
   a) prokaryotic
   b) eukaryotic

7. Unicellular organisms within the kingdom protista are:
   a) prokaryotic
   b) eukaryotic

8. What are two differences between prokaryotic and eukaryotic cells that are visible through a light microscope?

_____
_____
_____
_____
_____
_____

Unit 5: Origins

Contents

Source Cartoon: Microbes Rising   56

Cartoon Profile   57

Background   58

Study Questions   59

Visual Exercise   60

Graphic Organizer   61

Unit 6: Temperature Tolerance

Contents

Source Cartoon: Heated Microbes   62

Cartoon Profile   63

Background   64

Study Questions   65

Visual Exercise   66

Dr. Birdley Investigates: Microbes Rising

Name: _____
Class: _____ Date: _____

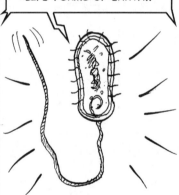
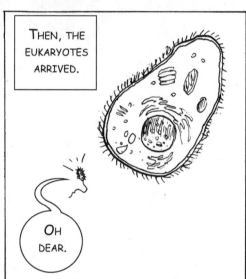

©Incentive Publications, Inc., Nashville, TN
©Nevin Katz

Dr. Birdley Teaches Science – Classifying Cells

Microbes Rising

Early Life

Objectives

To compare the eukaryotes with prokaryotes during earth's early history.

Background

Birdley reads a story describing how the prokaryotes were in the process of occupying every possible niche on earth. The eukaryotes, which are larger and more complex, then arrive.

Main Ideas

1. Bacteria (prokaryotes) were the first organisms to live on earth.
2. Prokaryotes existed 3.5 billion years ago.
3. Eukaryotes first emerged 2.5 billion years ago.
4. Eukaryotic cells were bigger and more complex.
5. Eukaryotes were generally higher on the food chains than prokaryotes.

Vocabulary

prokaryotes   nucleus
eukaryotes   bacteria
organelles   protists

Characters

Dr. Birdley, his son Iggy, prokaryotes, prokaryote leader, one-celled eukaryotes

Teacher's Note

The most recent findings suggest that the *first* prokaryotes existed about 3.8 billion years ago.

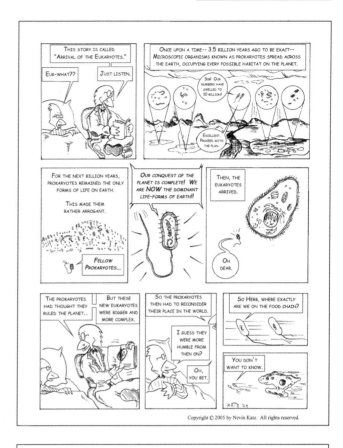

Questions for Discussion

Before Reading:

1. When do you think life began on earth?
2. What do you think were the first living things on earth?
3. Do you think they were multicellular or unicellular? Why?

After Reading?

1. How did life change for the prokaryotes?
2. How were the eukaryotes different?
3. Do you think our body cells prokaryotic or eukaryotic? Why?

©Incentive Publications, Inc., Nashville, TN

Dr. Birdley Teaches Science – Classifying Cells

PROKARYOTES AND EUKARYTOTES

NAME:_____
CLASS:_____ DATE:_____

Background: Microbes Rising

The purpose of this cartoon is to differentiate between prokaryotes and eukaryotes in the context of the earth's history. In Dr. Birdley's bedtime story, prokaryotes emerged on earth 3.5 billion years ago. They were primarily one-celled bacteria.

Over time, bacteria came to occupy all kinds of habitats, using a great range of energy resources. For example, some of these bacteria developed chlorophyll, which allowed them to use the sun's energy for making food. Through continued use of photosynthesis, these prokaryotes filled the earth with oxygen, helping to create an atmosphere that could support oxygen-breathing life.

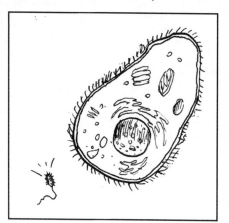

The prokaryotes were alone until 2.2 billion years ago, when eukaryotes arrived. The "euks" started off as unicellular, but eventually multicellular organisms arrived. Eukaryotic cells were generally bigger and more complex. This altered the ecological balance considerably. Suddenly the bacteria were living among predators that could devour them at any time!

Despite the new presence of eukaryotes, the prokaryotes would continue to thrive and evolve. Prokaryotes would capitalize on strengths such as a high rate of reproduction. They can also exchange DNA with each other, creating diversity within their populations. In a given environment, the strongest bacteria of this diverse group would survive, and continue to reproduce. This continuous natural selection has resulted in 3.8 billion years of evolution, resulting in prokaryotes being the most evolutionarily advanced type of life form on earth.

Prokaryotes have proven that they can find countless habitats, such as sulfur pools, doorknobs, all over the human body, and within our digestive tract. Watch out! They are everywhere.

---

1. How did bacteria create a suitable environment for living things?

_____

_____

2. What are two advantages that prokaryotes had over eukaryotes?

_____

_____

MICROBES RISING

NAME:_____
CLASS:_____ DATE:_____

 STUDY QUESTIONS

Directions: Answer the following questions to the best of your ability.

1. Based on the cartoon, what do you think a prokaryote is?
_____
_____

2. Based on the images and the story in the cartoon, what behaviors would you attribute to the prokaryotes?
_____
_____

3. List two reasons why the prokaryotes became "arrogant" (over-confident) in the story.
_____
_____

4. What was the name of the new group of organisms that arrived? How were they different from the prokaryotes?
_____
_____

5. Based on the information in the two final panels, how would the prokaryotes' ecosystem change after the arrival of the eukaryotes?
_____
_____

©Incentive Publications, Inc., Nashville, TN

Dr. Birdley Teaches Science – Classifying Cells

THE PROKARYOTES STRIKE BACK!

NAME:_____
CLASS:_____ DATE:_____

"ALRIGHT, PEOPLE! FILL IN THE BLANKS WITH THE WORDS IN THE BANK!!"

"DO IT NOW."

WORD BANK

DIVERSE	PROKARYOTES
EUKARYOTES	EXCHANGE
HABITATS (2)	SURVIVAL
RESISTANT	REPRODUCE
TEMPERATURES	POPULATIONS
SMALL	

---

1. ONE BILLION YEARS AGO, THE E_____ ARRIVED TO GREET THE P_____.

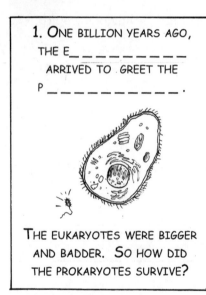

THE EUKARYOTES WERE BIGGER AND BADDER. SO HOW DID THE PROKARYOTES SURVIVE?

2. EVEN THOUGH THEY WERE S_____ AND SIMPLE, PROKARYOTES WERE STURDY LITTLE GUYS.

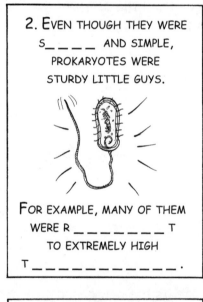

FOR EXAMPLE, MANY OF THEM WERE R_____ T TO EXTREMELY HIGH T_____.

3. AS A RESULT, "PROKS" COULD VENTURE INTO H_____ THAT "EUKS" WOULD NOT DARE TO ENTER.

"TOO HOT!!"

---

4. PROKARYOTES ARE ALSO ABLE TO E_____ BITS AND PIECES OF THEIR DNA.

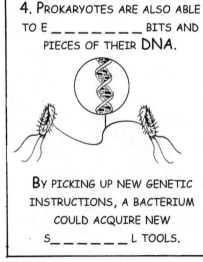

BY PICKING UP NEW GENETIC INSTRUCTIONS, A BACTERIUM COULD ACQUIRE NEW S_____ L TOOLS.

5. AS A RESULT OF DNA EXCHANGE, PROKARYOTES GREW INTO A D_____ GROUP OF CREATURES...

...THAT CAME TO OCCUPY COUNTLESS TYPES OF H_____ ON EARTH!

6. PROKARYOTES COULD ALSO R_____ AT A FASTER RATE THAN THE LARGER EUKARYOTES....

ALLOWING THEIR P_____S TO GROW QUICKLY OVER SHORT LENGTHS OF TIME.

DR. BIRDLEY INVESTIGATES — MICROBES RISING

NAME:_____
CLASS:_____ DATE:_____

 Graphic Organizer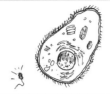

Fill in the table using information from the source cartoons "Microbes Rising" and "Two Types of Cells," as well as the visual exercise "Prokaryotes Strike Back."

	Prokaryotes	Eukaryotes
Size		
Internal Structure		
Strengths and Advantages		
Types of organisms within this category		

©Incentive Publications, Inc., Nashville, TN

Dr. Birdley Teaches Science – Classifying Cells

Heated Microbes

Name:_____
Class:_____ Date:_____

Heated Microbes

Extremophiles

Objectives

1. To illustrate an experimental setup involving microbes.

2. To provide an example of how some prokaryotes have an advantage over eukaryotes.

Synopsis

Christina is heating a beaker full of microbes to eliminate those that cannot tolerate high temperatures. A paramecium is about to devour some smaller bacteria. The temperature of water gets hotter, the paramecium dies at 37°C, and the smaller bacteria survive.

Main Ideas

1. Although eukaryotes are bigger and more complex, prokaryotes generally can survive higher temperatures.

2. The experimental setup involves a beaker, hot plate, thermometer, clamp, and ringstand.

3. Like all organisms, the paramecium needs a temperature range that allows it to maintain homeostasis and adequately regulate its internal conditions.

4. The bacteria are the prokaryotes.

Vocabulary

beaker	clamp	eukaryotes
hot plate	homeostasis	protist
ringstand	prokaryotes	paramecium

Characters

Christine, Anthony, the Paramecium, bacteria

Questions for Discussion

Before Reading:

1. What strengths do the single-celled eukaryotes (protists) have over the single-celled prokaryotes (bacteria)?

2. How could you heat a liquid and measure its temperature at the same time? List the equipment and draw the setup.

After Reading:

1. What advantage do some prokaryotes have over eukaryotes?

2. How did Christina isolate the most temperature-resistant bacteria?

©Incentive Publications, Inc., Nashville, TN

Dr. Birdley Teaches Science – Classifying Cells

Prokaryotes and Eukaryotes

NAME:_____
CLASS:_____ DATE:_____

Background: Heated Microbes

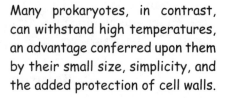

The purpose of this cartoon is to point out that prokaryotes, a classification which includes all bacteria, can withstand a wider range of temperatures than their eukaryotic buddies.

Every microorganism has an upper thermal death point: the temperature that is too hot for it to survive. Many prokaryotes have higher upper thermal death points than eukaryotes. Christina aims to exploit this difference by heating a beaker of water that is full of microbes. The temperature of the beaker is being monitored by the thermometer, and indicated in the cartoon panels.

Within the beaker, a paramecium is attacking a flock of bacteria. It boasts that eukaryotes are the new rulers of the microbial world. However, when the temperature reaches 37° Celsius, the paramecium dies. The bacteria remain, commenting on the inability of eukaryotic microbes to withstand high temperatures.

The upper thermal death point of the paramecium is in fact 37° Celsius. At this point, its membrane destabilizes, it can no longer produce cell parts, and its cellular metabolism breaks down. In general, eukaryotic microbes are unable to tolerate extremely high temperatures.

Many prokaryotes, in contrast, can withstand high temperatures, an advantage conferred upon them by their small size, simplicity, and the added protection of cell walls.

Tolerance for high temperatures allows some prokaryotes to occupy niches where eukaryotes could not survive. For example, some types of archaebacteria live in ocean floor vents, where temperatures are above boiling. Other bacteria live in the steaming sulfur pools of Yellowstone Park. So be careful next time you assume eukaryotes have the total advantage. The prokaryotes' ability to withstand high temperatures is unmatched!

___

1. Explain the experimental setup in your own words.

___

___

2. What advantage do prokaryotes have over eukaryotes?

___

___

 HEATED MICROBES

NAME:_____
CLASS:_____ DATE:_____

STUDY QUESTIONS

Directions: Answer the following questions to the best of your ability.

   1. Describe Christina's experimental setup in the cartoon.
_____
_____

   2. What is Christina's goal in heating the beaker? Is it ethical?
_____
_____

   3. List the two major types of living things in the beaker. What's happening?
_____
_____

   4. Give two reasons why the *Paramecium Caudatum* thinks it is superior to the bacteria.
_____
_____

   5. Are prokaryotes inferior to eukaryotes in every conceivable way? Why or why not?
_____
_____

Christina's Equipment

NAME:_____
CLASS:_____ DATE:_____

Useful Words

WATER	MEASURE
INCREASE	CONTAIN
TEMPERATURE	IN PLACE
HOLD	KEEP
SUSPEND	MICROBES

1. Hot Plate _____
_____
_____
_____

2. Beaker _____
_____
_____
_____

3. Thermometer _____
_____
_____
_____

4. Ringstand _____
_____
_____
_____

5. Clamp _____
_____
_____
_____

Unit 7: Cell Scientists

Contents

Source Cartoons: Hooked on Cells   68
　　　　　　　　　A Small World   69

Cartoon Profiles (2)   70 and 71

Study Questions (2)   72 and 73

Source Cartoons: Schleiden and the Plants   74
　　　　　　　　　A Meeting of the Minds   75

Cartoon Profiles (2)   76 and 77

Study Questions (2)   78 and 79

Source Cartoons: Constructing a Theory   80
　　　　　　　　　Rudolph Virchow   81

Cartoon Profiles (2)   82 and 83

Study Questions (2)   84 and 85

Background   86

Visual Exercise   87

HOOKED ON CELLS

Name:_____
Class:_____ Date:_____

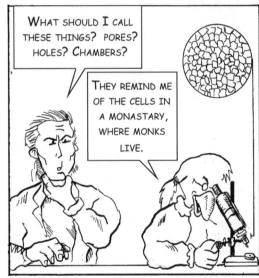

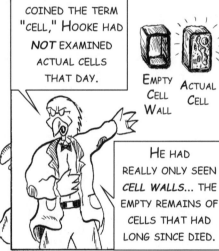

©Incentive Publications, Inc., Nashville, TN
©Nevin Katz

Dr. Birdley Teaches Science – Classifying Cells

Hooked on Cells

History of Biology

Objectives

To illustrate how Hooke first discovered cells and coined their name.

Synopsis

Dr. Birdley tells the story of Robert Hooke, who finds that a slice of cork is made of tiny compartments. After naming them cells, Hooke goes on to examine more specimens and publish his findings.

Main Ideas

1. In 1655, Robert Hooke discovered that a slice of cork was made up of tiny compartments.
2. He named these compartments cells, because they reminded him of the cells in a monastery.
3. The cork specimen actually contained cell walls, the empty remains of cells that had long since died.
4. Hooke went on to examine and draw more specimens, which included live cells.
5. He published his findings in the book, *Micrographia*.

Vocabulary

cork   compartments   monastery
cells   cell walls   *Micrographia*

Characters

Dr. Birdley, Hooke, Louis Gosling, four members of the 17th century scientific community, Louis Gosling's confidante.

Questions for Discussion

Before Reading:

1. At what point in history were cells discovered? Why do you pick this point?
2. What do you think were the first specimens that people found to be made out of cells?

After Reading:

1. What is Hooke's place in the history of science?
2. Where does the term "cell" come from?
3. How was Hooke's first specimen in the cartoon different from actual cells?

©Incentive Publications, Inc., Nashville, TN   Dr. Birdley Teaches Science – Classifying Cells

A Small World

History of Biology

Objectives

To portray Leeuwenhoek's initial glimpse into the world of the microbes, and highlight some of his major contributions to biology.

Synopsis

Dean Owelle tells the story of Antony Van Leeuwenhoek, a man with little formal science training who discovered the first one-celled microorganisms using his one-lens microscope.

This discovery not only helped to confirm the existence of cells, but also led him to discover additional cell types. Leeuwenhoek's research helped lay the foundations for later microbiology research.

Main Ideas

1. Leeuwenhoek lived in Delft, Holland, in 1673, where he worked as a dry-goods dealer.

2. Leeuwenhoek enjoyed grinding lenses and building microscopes.

3. While examining a drop of rainwater through his microscope, Leeuwenhoek discovered one celled microorganisms.

4. Leeuwenhoek went on to discover bacteria, protists, blood cells, and sperm cells.

Vocabulary

microorganism   unicellular organism

Characters

Dean Owelle, Leeuwenhoek, Microbes, Dr. Birdley

Questions for Discussion

Before Reading:

1. What are the qualities of a great scientist?

2. Is it possible to be a scientist with little formal science training? Explain.

After Reading:

1. What was Leeuwenhoek's first major discovery?

2. From what you can see, how is his microscope different from other microscopes you have seen?

3. If you were a biologist back in his day, how might you have shaped your research in response to Leeuwenhoek's findings? Give an example.

©Incentive Publications, Inc., Nashville, TN

Dr. Birdley Teaches Science – Classifying Cells

HOOKED ON CELLS

NAME:_____
CLASS:_____ DATE:_____

STUDY QUESTIONS

Directions: Answer the following questions to the best of your ability.

1. What was the first piece of evidence Hooke found that suggested cells existed? Where did this evidence come from?
_____
_____

2. Why was the term "cell" chosen?
_____
_____

3. How were Hooke's initial findings different from actual cells?
_____
_____

4. How fid Hooke communicated his findings to the scientific community? If you were a biologist at this time, how would you respond to these findings?
_____
_____

5. Discuss Hooke's microscope. Write about its importance in history, and how it is different from today's microscopes.
_____
_____

A SMALL WORLD

NAME:_____
CLASS:_____ DATE:_____

 STUDY QUESTIONS

Directions: Answer the following questions to the best of your ability.

1. What exactly did Leeuwenhoek discover? Why was it significant in the overall field of biology at that time?
_____
_____

2. Leeuwenhoek enjoyed making excellent lenses and microscopes. How do you think this activity helped him in his scientific pursuits?
_____
_____

3. Contrast Leeuwenhoek's microscope with the traditional compound microscope Point out at least three differences.
_____
_____

4. Why do you think Leeuwenhoek was confused by the absence of heads and tails on his specimens?
_____
_____

5. How were Leeuwenhoek's other discoveries similar to each other? What conclusions or generalizations can you draw from his discoveries?
_____
_____

©Incentive Publications, Inc., Nashville, TN

Dr. Birdley Teaches Science – Classifying Cells

SCHLEIDEN AND THE PLANTS

Name:_____
Class:_____ Date:_____

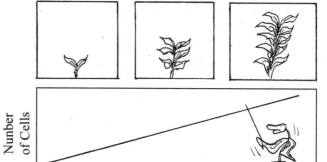

A Meeting of the Minds

Name:_____
Class:_____ Date:_____

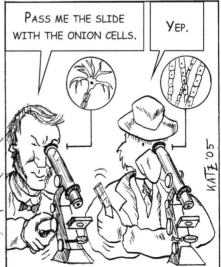

Cartoon Profile: The Cell Theory

Objectives

To illustrate how Schleiden's studies helped to lay the groundwork for the cell theory.

Synopsis

Phyll explains how Schleiden discovered the importance of cells as the basis for structure and function in all plants.

Main Ideas

1. Schleiden discovered that all plants were made up of cells.
2. Schleiden found that all plant parts were either made of or produced by cells.
3. Schleiden found that as plants grew, the number of cells in the plant grew as well.
4. After Schleiden had discovered that all plants were made of cells, another scientist (Theodor Schwann) would investigate whether the same was true for animals.

Vocabulary

cell division   mitosis

Characters

Professor Brockley, Phyll, Schleiden

Teacher's Note

This is the beginning of a four-page cartoon narrative on how the cell theory was initially formulated. The subsequent comics, *Meeting of the Minds, Constructing a Theory,* and *Rudolph Virchow* continue to tell about the scientists who came up with the cell theory.

Questions for Discussion

Before Reading:

1. Are plants easier to study than animals? Why or why not?
2. What do you think cells do that causes plant growth?

After Reading:

1. Why is Phyll (the plant) offended?
2. What is Schleiden's overall role in the history of science?
3. What are three of Schleiden's discoveries?
4. How are Schleiden's discoveries different?

©Incentive Publications, Inc., Nashville, TN

Dr. Birdley Teaches Science – Classifying Cells

Cartoon Profile: A Meeting of the Minds

Objective

To understand how the work of Schwann and Schleiden led to the initial ideas for the cell theory.

Synopsis

Brockley begins the story of Theodor Schwann, who compares plant and animal tissues with his colleague, Matthias Schleiden. After realizing that both have cell-like qualities, he sets out to prove that both plants and animals are made of cells.

Main Ideas

1. Schwann initially realized that animal tissues were made up of tiny compartments.

2. Schwann told this to Schleiden, and they went on to examine and compare plant and animal tissues in the lab.

3. After the initial examination, it was apparent that both plant and animal tissues were made up of cells.

4. Schwann formed a hypothesis that both plants and animals were made up of cells, and set out to prove it.

Vocabulary

cell   tissue   hypothesis
nucleus   membrane   theory

Characters

Brockley, Phyll, Jacob Cardinal, Theodor Schwann, Matthias Schleiden

Questions for Discussion

Before Reading:

1. What types of experiments would help you prove that all living things are made up of cells?

2. What materials would you need to run these experiments?

After Reading:

1. What was Schwann's hypothesis by the end of the cartoon?

2. What inspired him to come up with this hypothesis?

3. What knowledge did Matthias Schleiden provide?

©Incentive Publications, Inc., Nashville, TN

Dr. Birdley Teaches Science – Classifying Cells

SCHLEIDEN AND THE PLANTS

NAME:_____
CLASS:_____ DATE:_____

 STUDY QUESTIONS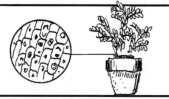

Directions: Answer the following questions to the best of your ability.

 1. Explain the three conclusions that Schleiden made as a result of his studies.
_____
_____

 2. Compare and contrast Schleiden's discoveries and conclusions with those of Robert Hooke.
_____
_____

 3. Compare and contrast Schleiden's discoveries and conclusions with those of Leeuwenhoek.
_____
_____

 4. Your plant grows two inches. Based on Schleiden's discoveries, what can you conclude about the cells in the plant?
_____
_____

 5. How were Schleiden's discoveries important to the overall progress of cell research? What next steps would further prove the importance of cells?
_____
_____

©Incentive Publications, Inc., Nashville, TN

Dr. Birdley Teaches Science – Classifying Cells

A MEETING OF THE MINDS

NAME:_____
CLASS:_____ DATE:_____

STUDY QUESTIONS

Directions: Answer the following questions to the best of your ability.

1. What was the topic of Schwann's and Schleiden's after-dinner conversation?

   _____

   _____

2. What activity was the result of this conversation? Why?

   _____

   _____

3. What was Schwann's hypothesis about cells?

   _____

   _____

4. How did Schwann begin to test his hypothesis?

   _____

   _____

5. Compare Schwann's hypothesis with Schleiden's statements in the previous cartoon.

   _____

   _____

©Incentive Publications, Inc., Nashville, TN

Dr. Birdley Teaches Science – Classifying Cells

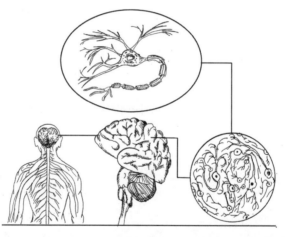

Cartoon Profile: Constructing a Theory

Objectives

To introduce Schwann's initial cell theory and point out the error in his third statement.

Synopsis

After collecting sufficient evidence, Schwann writes the first cell theory in 1838. Twenty years later, the third part of his theory is proven incorrect.

Main Ideas

1. Theodor Schwann wrote the first version of the cell theory in 1838 based on his findings.

2. The theory correctly stated that cells were the smallest units of life, and that all living things were made of cells.

3. The theory incorrectly stated that new cells self-assembled in between pre-existing cells.

4. This error was pointed out in 1858.

Vocabulary

cells	theory	hypothesis
mitosis	fact	species
slides	opinion	tissue

Characters

Brockley, Schwann, Jacob Cardinal

Teacher's Note

Although Theodor Schwann wrote the initial cell theory, both he and Matthias Schleiden are credited with making comparable contributions to its creation.

Questions for Discussion

Before Reading:

1. What type of evidence would Schwann need before writing the cell theory?

2. What do you already know about the cell theory?

After Reading:

1. Which statement of his cell theory was deemed incorrect?

2. How would you rewrite this statement to make it correct?

3. What type of discoveries do you think led people to disprove this third statement?

©Incentive Publications, Inc., Nashville, TN

Dr. Birdley Teaches Science – Classifying Cells

CARTOON PROFILE: RUDOLPH VIRCHOW

Objectives

To explain how and why Schwann's third statement in the cell theory was revised.

Synopsis

A search for more information on Schwann's critics leads Jacob to Rudolph Virchow, a scientist who has correctly discovered how new cells form.

Main Ideas

1. Schwann's third statement about cells spontaneously forming in between other cells is incorrect because cells cannot build themselves.

2. Twenty years after Schwann's cell theory was written, Virchow had found that cells reproduce by dividing in two.

3. As a result, Virchow could claim that all cells come from directly from pre-existing cells.

4. Virchow's discoveries led to a revision in the third statement of the cell theory.

Vocabulary

cells   theory   hypothesis
division   fact   species
slides   opinion   tissue

Characters

Jacob Cardinal, Rudolph Virchow, Schwann, and miscellaneous scientists

Questions for Discussion

Before Reading:

1. Which part of Schwann's theory was incorrect?

2. Can theories change? Why or why not?

After Reading:

1. How is Virchow's statement different from Schwann's third statement?

2. If you were Schwann, how would you feel about your work being revised?

3. Can you think of any other universal statements about cells that could be incorporated into the cell theory?

©Incentive Publications, Inc., Nashville, TN

Dr. Birdley Teaches Science – Classifying Cells

CONSTRUCTING A THEORY

NAME:_____
CLASS:_____ DATE:_____

 STUDY QUESTIONS

Directions: Answer the following questions to the best of your ability.

1. What observations did Schwann make before writing the cell theory?
_____
_____

2. List the system, organ, tissue, and cell that are pictured under the first statement (second frame, middle row, page 80).
_____
_____

3. List three types of organisms under the second statement (third frame).
_____
_____

4. Which of the statements is incorrect?
_____
_____

5. How do you think new cells form?
_____
_____

RUDOLPH VIRCHOW

NAME:_____
CLASS:_____ DATE:_____

STUDY QUESTIONS

Directions: Answer the following questions to the best of your ability.

1. Why did Jacob want to go see Rudolph Virchow?
_____
_____

2. What was Virchow's argument against Schwann's third statement?
_____
_____

3. What was Virchow's explanation for how new cells formed?
_____
_____

4. How was the cell theory revised?
_____
_____

5. What do you think Virchow needed to obtain in order to revise the cell theory?
_____
_____

THE CELL THEORY

NAME:_____
CLASS:_____DATE:_____

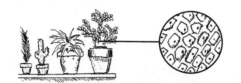 **BACKGROUND**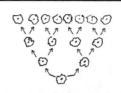

A theory is different from a hypothesis because it is supported by scientific evidence The cell theory is a good example of this because although it has been revised over time, experimental findings have continued to support its central ideas.

In the 19th century, there were three scientists who helped develop the cell theory: Theodor Schwann, Mathias Schleiden, and Rudolph Virchow. In 1838, Schleiden found that every plant he examined was made of cells. In 1839, Schwann had found that all the animals he had been looking at were made of cells, and then learned of Schleiden's results.

After a great deal of research, Schwann wrote the first cell theory, which correctly stated that all living things are made of cells, and that cells are the smallest unit of life. His third statement on cell reproduction, which stated that cells formed spontaneously like crystals, was incorrect. In 1859, Rudolph Virchow revised this idea with the statement that all living things come from pre-existing cells.

Today, three more statements have been added:
- All cells contain hereditary information that is passed on during cell division.
- All cells are basically the same in terms of chemical composition.
- All energy flow of life occurs within cells.

Directions: Name two to three scientists who helped develop the cell theory. Explain their contributions.

_____
_____
_____
_____
_____
_____
_____

©Incentive Publications, Inc., Nashville, TN

Dr. Birdley Teaches Science – Classifying Cells

Answer Key

Answer Key:

Open-ended questions give science students a chance to make connections between new concepts and their personal experiences, as well as to express their answers in different ways. These questions are one important way to differentiate instruction and address different learning styles. However, open-ended questions do not have a single correct answer, so please read student responses carefully to make sure they are not forming misconceptions. One possible correct answer is always given in this answer key

Page 17: Birds on the Run Study Questions
1. Cardiac cells keep the heart beating properly by working together to control their rhythmic contractions.
2. The force generated by the heart's pumping enables red blood cells to circulate around the body.
3. The critical protein molecule located on red blood cells is hemoglobin. It is important because it captures oxygen molecules and drops them off when it reaches body tissues.
4. Answers will vary. Nerve cells send and receive signals throughout the body, and control thought as well as automatic and conscious movement. Muscle cells use energy to produce mechanical motion.
5. Answers will vary. Muscle cells need oxygen and nutrients to produce motion.

Page 18: The T-Cell Background Exercises
T-Cells are similar to nerve cells in that their "jobs" rely on communication. Whereas T-cells communicate with each other about "non-self" invaders that they must fight off, nerve cells communicate in order to control the senses, movement, thought, and other body functions involving voluntary and involuntary control.

T-cells are similar to red blood cells in that they are both present in the bloodstream. Whereas red blood cells deliver oxygen, T-cells fight off viruses and bacteria.

Page 19: Living Body Cells Visual Exercise – see p. 91

Page 20: Body Cell Types Visual Exercise – see p. 91

Page 22: Connecting Terms Panel Review
1. The heart is made up of cardiac muscle cells. The contractions and relaxations of cardiac muscle cells cause the heart to beat.
2. The heart pumps red blood cells through the circulatory system.
3. Blood cells deliver oxygen to muscle cells.
4. Nerve cells transmit signals from the brain to muscle cells, which "tell" them when to contract and relax.

Page 23: Birds on the Run Graphic Organizer – see p. 93

Page 24: Unit 1 Quiz: Body Cells

1. b
2. a
3. c
4. a
5. c
6. d
7. Answers will vary. Nerve cells and striated muscle cells are both similar in that they both help enable the organism to respond to its environment. They are different because only the muscle cells create mechanical motion, while the nerve cells only process information.

Page 28: If You Were a Cell Background Exercise
1. One advantage of being a body cell in a multicellular organism is that you work with other cells and are protected to some extent by the organism you are in (for as long as you are useful!) One disadvantage is that you do not have much independence.
2. One advantage of being a unicellular organism is that there is more independence than being a body cell in a multicellular organism. One disadvantage is that there could be more danger, depending on the environment you find yourself in.

Page 29: If You Were A Cell Study Questions
1. Answers will vary. Bacteria and one-celled protists would be unicellular organisms, whereas body cells of animals, plant cells, and fungi cells would be part of a multicellular organism.
2. Nerve cells make up the tissue that the brain is composed of.
3. Answers will vary. The cartoon points out that unicellular organisms can live independently, but have to survive within ecosystems that contain predators.
4. Answers will vary. The cartoon points out that cells in multicellular organisms cooperate and work together, but are not able to exist by themselves.
5. Answers will vary, but may include a range of environments from the gut inside of mammals, to lakes, to undersea vents.

Page 30: Cell Story Starters Visual Exercise – see p. 91

Page 38: The Assistant / Kingdoms' Overview
1. Two multicellular organisms include a plant and an animal. Whereas an animal eats to obtain nutrients, a plant uses photosynthesis to make its own food. Most fungi are multicellular, and obtain nutrients by producing chemicals that break down organic matter.
2. Two unicellular organisms include bacteria and one-celled protists. Whereas a once-celled protist is relatively large and has one or more nuclei, a bacterium is smaller and does not have a nucleus.

Page 39: The Assistant / Kingdoms' Overview Study Questions
1. Multicellular organisms are made of many cells, but unicellular organisms are made of one cell.
2. Examples will vary. Three kingdoms that contain mainly multicellular living things include: animalia (rabbit), plantae (pine tree), and fungi (mushroom).
3. Two kingdoms that include both multicellular and unicellular organisms include fungi (mold, yeast) and protists (spirogyra, paramecium).
4. Answers will vary. Spirogyra is different from plants because it is generally simpler in structure, lacking leaves, roots, and a stem.
5. Answers will vary. The kingdoms that were not established until after the invention of the microscope were eubacteria, archaebacteria, and protista, because the majority of species within those kingdoms are too small to be seen with the unaided eye.

Page 40: Classify It Visual Exercise – see p. 91

Page 41: Classify It Visual Exercise, continued – see p. 92

Answer Key

Page 44: Unit 3 Quiz: Cells and the Kingdoms

1. c
2. a
3. b
4. a
5. a
6. Answers will vary. While both cells use sunlight, water, and carbon dioxide to produce oxygen, one is a self-sustaining organism while the other plays a specialized role within a larger organism.

Page 48: Two Types of Cells Background Exercise

1. Whereas eukaryotes have a nucleus and large, membrane-bound organelles, prokaryotes do not. Their DNA floats in the cytoplasm, and they only have small internal cell parts, such as ribosomes.
2. Under a light microscope, eukaryotes are generally bigger than prokaryotes. With the proper stain, there is a spot (the nucleus) that can be seen on eukaryotic cells, which is not present in prokaryotic cells.

Page 49: Two Types of Cells Study Questions

1. The two major types of cells being discussed are prokaryotes and eukaryotes. The apparent differences are that eukaryotes are bigger and have spots called nuclei.
2. While the eukaryotic cell has large organelles and its DNA is contained in the nucleus. In contrast, a prokaryotic cell only has small organelles and its DNA floats freely in the cytoplasm.
3. Prokaryotes probably have less DNA than eukaryotes because they are simpler. In addition, their DNA is not protected in a nucleus.
4. Four kingdoms that include species with eukaryotic cells are animalia, plantae, protista, and fungi.
5. Answers will vary. The cell that spoke to Dr. Birdley is a prokaryote, and is offended because Dr. Birdley called them simple. Dr. Birdley may have said this because compared to eukaryotes, prokaryotes are structurally less complicated.

Page 50: Cell Type Identification Visual Exercise – see p. 92

Page 51: Cell Type Identification, continued Visual Exercise – see p. 92

Page 54: Unit 4 Quiz: Two Types of Cells

1. a
2. c
3. c
4. b
5. b
6. a
7. b
8. Answers will vary. Under a light microscope, using the correct staining can show you that eukaryotic cells are relatively larger and have a nucleus, whereas prokaryotic cells are relatively smaller and do not have a nucleus.

Page 58: Microbes Rising Background Exercises

1. Bacteria created a suitable environment by filling the world with oxygen—a byproduct of their photosynthetic activities. The presence of oxygen "paved the way" for oxygen-breathing organisms, such as animals.
2. Some prokaryotes can tolerate higher temperatures than the most "temperature-hardy" eukaryotes. Prokaryotes have a higher rate of reproduction, and can exchange DNA.

Page 59: Microbes Rising Study Exercises

1. Answers will vary. A prokaryote is a microorganism that was one of the first life-forms to appear on earth.
2. Answers will vary. The prokaryotes appear to be one-celled organisms. They are different in terms of their various shapes. They produced oxygen, reproduced rapidly, exchanged DNA to create diversity, survived in many diverse habitats, and produced their own food.
3. The prokaryotes became arrogant because they were becoming more numerous, and seemed to be the only living things on the planet.
4. The new organisms that arrived were the eukaryotes. They were bigger than the prokaryotes and more complex.
5. The prokaryotes ecosystem would now include eukaryotes, some of which would be their predators. As a result, prokaryotes would not be as high up on the food chain.

Page 60: The Prokaryotes Strike Back! Visual Exercise – see p. 92

Page 61: Microbes Rising Visual Exercise – see p. 93

Page 64: Heated Microbes Background Exercises

1. In the experiment, the hot plate is heating a beaker, which contains water filled with microorganisms. (Christina will examine a sample of the water under a microscope at one point.)
2. Some prokaryotes are able to tolerate higher temperatures than eukaryotes.

Page 65: Heated Microbes Study Questions

1. In Christina's experimental setup, there is a beaker that is being heated by a hot plate below. Within the beaker is water, which contains microorganisms.
2. Answers will vary. Christina's goal is to heat the beaker to a point where only some of the microorganisms will survive. Whether its ethical depends on how much you care about microbes!
3. The two major types of living things are prokaryotes (bacteria) and a eukaryote (paramecium.) The paramecium is threatening to eat the bacteria.
4. The paramecium may think its superior because it is bigger, more complex, and occupies a higher place on the food chain.
5. Prokaryotes are not inferior to eukaryotes because in this case, they can tolerate higher temperatures.

Page 66: Christina's Equipment Visual Exercise – see p. 93

Page 72: Hooked on Cells Study Questions

1. Hooke's first piece of evidence was a slice of cork that came from the bark of a tree.
2. The term "cell" was chosen because the compartments reminded Hooke of the rooms, or cells, in a monastery.
3. Hooke found cell walls, which were the empty remains of cells that had died.
4. Hooke communicated his findings to others by publishing his work in the book *Micrographia.*
5. Hooke's microscope looks lesson complex, with less objectives and probably lower magnification. It was important because it enabled Hooke to discover cells.

©Incentive Publications, Inc., Nashville, TN

Dr. Birdley Teaches Science – Classifying Cells

Answer Key

Page 73: A Small World Study Questions
1. Leeuwenhoek discovered one-celled organisms, as well as a variety of other cell types. This was significant because few cell types had been discovered at that point in history.
2. Leeuwenhoek's lenses could have been used in his microscopes, enabling him to see his specimens.
3. In contrast to the traditional compound microscope, Leeuwnehoek's microscope had one lens, it had a tip for placing specimens instead of a stage for mounting slides, and it did not have as many magnification settings.
4. Answers will vary. Leeuwenhoek was confused about the specimens not having heads or tails because only moving living things he had seen before were animals.
5. Leeuwenhoek's additional discoveries were similar in that they all involved examining various types of cells. Students' conclusions will vary.

Page 78: Schleiden and the Plants Study Questions
1. Schleiden concluded that all plants were made of cells, all plant parts were made of or produced from cells, and that plant growth was due to the production of new cells.
2. Whereas Hooke was the first to discover cells, Schleiden was more concerned with discovering the importance of cells.
3. Whereas Leeuwenhoek examined microbes, body cells, and germ cells, Schleiden examined plant cells.
4. If a plant grows two inches, more cells are being produced.
5. Schleiden's work helped to establish the importance of cells in all living things. Further studies would determine if all animals were made of cells.

Page 79: Meeting of the Minds Study Questions
1. Schleiden and Schwann were discussing how the plants and animals they were looking at were all made up of cells.
2. They verified the idea that all living things were made of cells by examining slides of animal and plant tissue.
3. Schwann's hypothesis was that all living things were made up of cells.
4. Schwann began to test his hypothesis by examining more tissue samples of living things.
5. Answers will vary. Whereas Schwann was trying to prove the universality of cells in all living things, Schleiden's statements were limited to plants.

Page 84: Constructing a Theory Study Questions
1. Before proposing his theory, Schwann observed evidence that all living things were made of cells, and that cells were the smallest units of life.
2. The nervous system, brain, nerve cell are pictured.
3. Three types of organisms are plants, animals, and fungi.
4. The third statement about cell reproduction is incorrect.
5. Answers will vary.

Page 85: Rudolph Virchow Study Questions
1. Jacob wanted to go to Rudolph Virchow to learn about the flaw in Schwann's cell theory.
2. Virchow's argument against Schwann's third statement was that cells could not construct themselves. Schwann had originally stated that cells formed spontaneously in between other cells.
3. Virchow stated that cells gave rise to new cells through cell division.
4. The third statement of the cell theory was revised to say that all cells came from pre-existing cells.
5. Virchow needed to obtain evidence for cell division in all types of organisms in order to revise his theory.

Page 86: Cell Theory Background Exercises
Matthias Schleiden discovered that all plants are made of cells.

Theodor Schwann discovered that all animals were made of cells, and developed the first cell theory. (All living things are made of cells; cells are the smallest units of life; cells form spontaneously in between pre-existing cells.) His last statement was in correct.

Rudolph Virchow stated that all cells come from pre-existing cells

Page 19 Living Body Cells

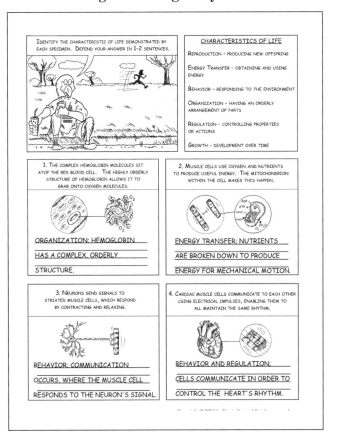

Page 20 Body Cell Types

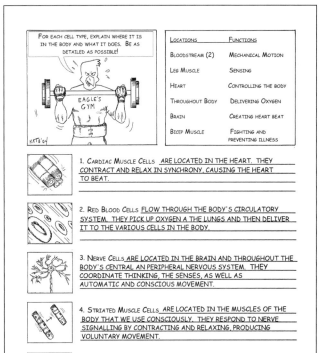

Page 30 Cell Story Starters

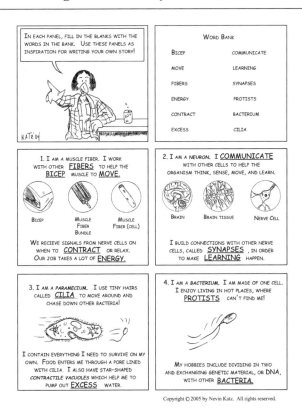

Page 40 Classify It!

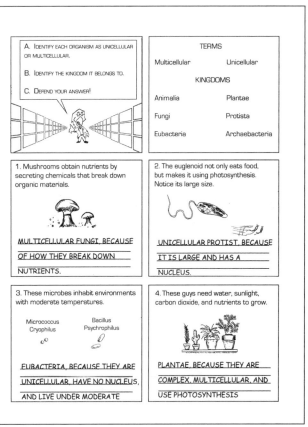

Page 41 Classify It! page 2

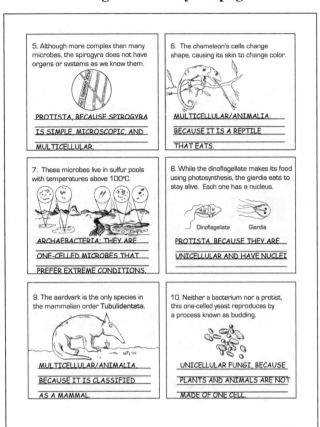

Page 50 Cell Type Identification

Page 51 Cell Type Identification, part 2

Page 60 Prokaryotes Strike Back!

Page 66 Christina's Experiment

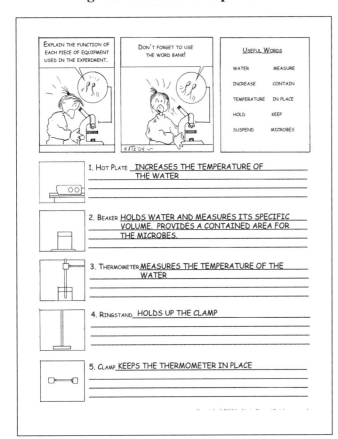

GRAPHIC ORGANIZERS

Birds on the Run (p. 23)

Cell Type & Location	Function
Cardiac Muscle Cell, Heart	Generates heart beat
Red Blood Cell, Bloodstream	Delivers Oxygen to cells and tissues
Nerve Cell, Neuromuscular Junction or brain	Coordinates movement
Muscle Cell, Leg Muscle	Mechanical Motion
T-Cell	fights off sickness-causing microbes

Microbes Rising (p. 61)

	Prokaryotes	Eukaryotes
Size	relatively small	relatively large
Internal Structure	cytoplasm and dna	cytoplasm, nucleus, other organelles
Strengths and Advantages	reproduces faster, can exchange DNA, higher temperature tolerance	bigger, more complex
Types of organisms within this category	bacteria	protists fungi animals plants